Table of Conte

Science Action Labs

Dear Teacher or Parent,

The spirit of Sir Isaac Newton will be with you and your students in this book. Newton loved science, math and experimenting. He explained the laws of gravity. He demonstrated the nature of light. He discovered how planets stay in orbit around our sun.

Sound & Light can help your students in many ways. Choose some activities to brighten up your class demonstrations. Some **sound and light units** can be converted to hands-on lab activities for the entire class. Some can be developed into student projects or reports. Every class has a few students with a special zest for science. Encourage them to pursue some science activities on their own.

Enjoy these science experiments as much as Newton would have. They are designed to make your students **think**. Thinking and solving problems are what science is all about. Each **sound and light unit** encourages thought. Students are often asked to come up with their best and most reasonable guess as to what will happen. Scientists call this type of guess a **hypothesis**. They are told how to assemble the materials necessary to actually try out each activity. Scientists call this **experimenting**.

Don't expect your experiments to always prove the hypothesis right. These science activities have been picked to challenge students' thinking abilities.

All the activities in **Sound & Light** are based upon science principles. That is why Sir Isaac Newton has been used as a guide through the pages of this book. Newton will help your students think about, build and experiment with these activities. Newton will be with them in every activity to advise, encourage and praise their efforts.

The answers you will need are on page 64. You will also find some science facts that will help your students understand what happened.

Here are some suggestions to help you succeed:

1. **Observe carefully.**
2. **Follow directions.**
3. **Measure carefully.**
4. **Hypothesize intelligently.**
5. **Experiment safely.**
6. **Keep experimenting until you succeed.**

Sincerely,

Ed

Ed Shevick

SOUND & LIGHT

Investigations in Sound & Light

Science Action Labs

Written by Edward Shevick

Illustrated by Leo Abbett

Teaching & Learning Company

1204 Buchanan St., P.O. Box 10

Carthage, IL 62321-0010

This book belongs to

_____

The activity portrayed on the front cover is described on page 13.

Cover design by Kelly Harl

Copyright © 2000, Teaching & Learning Company

ISBN No. 1-57310-209-1

Printing No. 987654321

Teaching & Learning Company
1204 Buchanan St., P.O. Box 10
Carthage, IL 62321-0010

Vibrating Sound

Newton Explains Sound

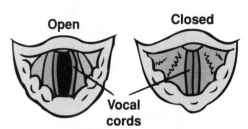

There are many kinds of sounds in the world. All sounds have one thing in common. All sounds are caused by something **vibrating**.

Your Vocal Cords

Place your fingers lightly on the place on your throat that is called your Adam's apple. It is your voice box or larynx. Your voice box has two thin muscles that vibrate. By controlling these two muscles, you can make all the sounds of speech.

1. Say "ah" and gently touch your Adam's apple. Say "ah" softly and then very loud.

What did you feel your Adam's apple doing? _____

Rubber Band Vibrations

1. Stretch a rubber band between your hands.

2. Have someone pluck the center of the rubber band.

What do you hear? _____

What do you see vibrating? _____

3. Try stretching your rubber band to make it longer and shorter. Pluck its center as before.

Describe the differences in sound and vibration between the long

and short rubber band. _____

_____

_____

Radio Cereal

1. Tie a light piece of thread to a small bit of O-shaped cereal.

2. Tape the thread to a radio so that the cereal is hanging freely, directly in front of the speaker.

Name _____

3. Turn your radio on **full blast** to a voice or music program.

Describe what happened to your cereal. _____

_____

What must the speaker be doing to cause the air and cereal to move? Hint: It's making sound and what is sound? _____

Identifying Sounds

You know that vocal cords and radio speakers vibrate to make sound. Sound travels from the vibrating object to air molecules. Air molecules bump into one another and cause your eardrums to vibrate. Your brain interprets what you are hearing.

1. Have everyone around you be especially quiet for three minutes.

Write down all the sounds that you hear in your inside and outside environment.

_____

_____

2. Compare what you heard to what others in your class heard.

What did you or they miss? _____

_____

Newton wants your class to have a sound contest. Try to come up with sounds that will stump your classmates. Here are the rules.

1. Your sound can be real or on tape.

2. Limit of three sounds per contestant.

3. You have three days to prepare your mystery sounds.

4. Forget the easy sounds. Make your sound different and hard to identify.

Humans and Sound

Sound plays a big part in your life. You learn through sound. You communicate through sound. A good tape or CD gives you pleasure.

6

Name _____

Sound can also be polluting. Loud sounds such as those given off by trucks, lawn mowers and airplanes can damage your ears.

1. Have someone blow up and then pop a balloon.

Did you find that a pleasant or unpleasant sound? _____

2. Obtain a radio and turn on some quiet music that you enjoy. Add numbers 1, 3, 5, 7, 9 and 11 in your head.

What is the answer? _____

3. Now turn your radio to a point in between stations to obtain static. Turn the volume up to its maximum. Add the numbers 2, 4, 6, 8, 10 and 12 in your head.

What is the answer? _____

Was it easier to add with pleasant music or loud, unpleasant static? Can you explain

why? _____

_____

Newton's Sound Riddle

Newton would like to finish this sound vibration unit with a riddle for you to solve. A tree falls to Earth deep within a forest. There is no human or animal around. Does the falling tree make a sound as it hits the Earth?

What is your answer to the riddle? _____

_____

Name _____

Sound Fun

Tines →

 ## The Vibrating Fork

You can make a table fork vibrate like a bell. Use two old all-metal forks, one for vibrating and the other as a hammer for striking.

1. Strike the tines of one fork with the handle of the other.

Can you observe the tines? _____

2. Strike the fork again and place it near your ear.

What did you hear? _____

3. Tie a light string to your fork. Tie a large button to the other end. Place the button on your ear and strike the fork. The string carried the sound to your ear.

Was the sound louder or softer? _____

 ## A Sound Experiment

1. Obtain an empty plastic tennis ball container.

2. Make a hole in the bottom about the size of a pea.

3. Cut off the neck of a balloon. Place the remainder of the balloon over the open end. You may have to use a rubber band to make it tight and secure.

4. Hold the end with the hole near your face. Tap the balloon end sharply.

What did you feel on your face? _____

_____

8

Name _____

Newton Knowledge: The puff of air you felt demonstrates how sound travels. You vibrated the rubber balloon. It pushed air molecules away. The molecules in the can pushed one another forward until some were forced out through the hole.

1. Have an adult light a candle set on a sturdy base.

2. Place the end with the hole 12" (30 cm) from and aimed at your lit candle.

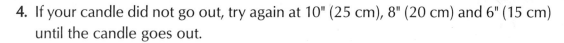

3. Pull on the balloon and release it.

4. If your candle did not go out, try again at 10" (25 cm), 8" (20 cm) and 6" (15 cm) until the candle goes out.

   How close did you have to get to blow out the candle? _____

Balloon Sound

Sound travels through air. It travels even faster through water or metal. That is because water and metal molecules are closer together than air molecules.

Remember your last swim. You actually heard sounds louder underwater. The sounds, however, were not clear. People in the early railroad days would put their ear to the metal tracks. They would hear the approaching trains before they could be heard through air sound waves.

You can amplify sounds with a balloon full of air.

1. Blow up a balloon the size of your head. Tie it so no air escapes.

2. Talk to a friend through the balloon. The loudness may surprise you. Could you pass sound through two or three balloons lined up? Try it.

Sound Travel

Sound travels only about 1100 feet (335 m) in one second. Light travels 186,000 miles (299,000 kilometers) in one second. If a sound is made at a reasonable distance, you see it **before** you hear it.

Let's try to demonstrate how sound is slower than light outdoors. You'll need a safe outdoor space at least 360 feet (109 m) in distance.

Name _____

1. Obtain a metal pail or pan and a hammer.

2. The students with the pail and hammer should move at least 360 feet (109 m) away from the class. This is slightly longer than a football field.

3. Hold the pail high so the distant students can see it.

4. Strike the pail vigorously with the hammer.

Was there some time between the students seeing the blow and hearing the sound?

_____

Newton Hint: At 360 feet (109 m) there is only about $\frac{1}{3}$ of a second between the arrival of the light and sound.

5. If possible, try this experiment at further distances.

Name _____

How Sound Travels

Newton Takes You to the Moon

Congratulations! You have just won tickets to a rock concert on the moon. Before you get too excited, there are some problems with a moon concert. How are you going to get there? What will you eat and drink?

There is another problem. There is no air on the moon. There is no sound without air. The sound you hear on Earth is transmitted to your ears by molecules of air.

Here is an **imaginary** experiment.

1. Place a glass jar or bowl over a ringing alarm clock. The sound will not be as loud, but you will still hear the ringing through the glass.

2. Now pump the air completely out of the glass jar.

   Why can't you hear the ringing in this imaginary experiment? _____

   _____

Air pumped out

Air Molecules in Motion

Air molecules are invisible. There are ways to show you how air molecules bump into one another to transmit sound vibrations.

1. Obtain four quarters.

2. Line them up as shown. The three quarters on the right side should be **touching**.

Name _____

3. Shove the fourth coin sharply against the two held coins. Repeat a few times.

What happened to the coin on the right? _____

_____

Newton Note: Your coins are like the invisible air molecule. Air molecules transmit the sound energy away from the source of vibrations.

Here's another way to demonstrate how air molecules transmit sound vibrations. Instead of coins, you will substitute people for air molecules.

1. Line up at least five people as shown. They should be in a row holding **both** outstretched arms on the shoulders in front of them.

2. Give a small shove to the first person in line.

What happened to the last person in line? _____

3. Give a **slightly** bigger shove as if you were making a loud sound.

What happened to the last person in line? _____

Slinky Sound

A Slinky™ spring toy can teach you a lot about sound waves. Imagine each coil in a Slinky™ to be an air molecule.

1. Obtain a Slinky™ toy.

2. Have a friend help you hold a Slinky™ on the floor. It stretches out to about 15 to 20 feet (4.5 to 6 m).

3. Send waves down the coil by pulling back and forth on **one** end.

Describe how the waves looked. Did the coils move forward one at a time or in

bunches? _____

Name _____

4. Send another wave down the Slinky™. This time look for the weaker wave that comes back toward you. This is like an echo in real sound.

Did your Slinky™ wave echo come back in bunches? _____

Newton Note: Real sound waves also move in bunches away from the vibrating source. The energy in your slinky coils only travels in one direction down the spring. Real sound waves travel in spheres in all directions from the source.

Sound Travels in All Directions

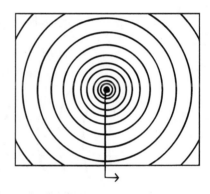

So far we have demonstrated how sound travels from air molecule to air molecule. Actual sound doesn't travel only in one direction. It travels in all directions from the source.

In some ways, sound waves are like water waves. Imagine a bathtub filled with water. You drop a rubber duck in the center of the tub. Waves will ripple out in all directions.

Here is a sound experiment you can do without a bathtub and a rubber duck.

1. Line up dominoes as shown in the diagram. They should go in four directions as shown. They should be close enough that one domino falls into another.

2. Drop a ball into the center of your dominoes.

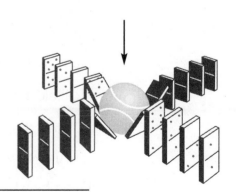

Describe what happened as if the dominoes were air molecules and the ball was a vibrating source of sound.

The water ripples and falling dominoes only show sound as traveling in two directions. Actual sound travels in all possible directions away from a vibrating source.

Think of yourself in the center of an onion. Start yelling. Imagine your voice traveling outward in **all** directions through the onion layers. Sound really moves in not two but three dimensions.

Sound Waves

Newton Explains Sound as an Energy Wave

Sound is a form of energy. Light is another form of energy. Energy can be transferred from one place to another in two ways. Energy can be transferred by particles or waves. Sound uses molecules of air as particles. Light can travel in waves through empty space. Light does not need particles.

Scientists find it convenient to think of sound as a wave. Waves can graphically show the characteristics of sound.

Looking at Waves

Here are two simple ways to observe waves similar to sound.

1. Hold a ruler over a table with about 8" (20 cm) hanging over the edge.

2. Hold the table end of the ruler firmly with your hand. Push the other end of the ruler down and then let go.

Describe what you saw and heard. _____

_____

3. Now place the ruler at the 6" (15 cm) mark. Repeat the experiment.

Describe the differences in sound and vibration. _____

_____

4. Obtain 20' (6.1 m) of string.

5. Tie one end to a doorknob.

14

Name _____

6. Pull the string toward you so that it is tightly stretched.

7. Move your hand up and down vigorously.

8. Have someone observe the string from the side.

Describe what happened to the string. _____

_____

9. Repeat the experiment holding the string even tighter and then looser.

Describe the differences in wave shape between the tight and loose strings.

_____

_____

Newton Special: There's a better way to observe sound waves. Hook up a clapper bell and some batteries. Tie a light string to the clapper. Turn the bell on and adjust the string's tension to observe waves.

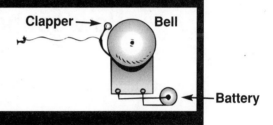

How Sounds Differ

Each sound has its own **pitch** and **frequency**. Your vocal cords can give off high and low frequencies. The frequency of a sound tells you how many waves go past a given point in one second. Let's assume the point is your eardrum. If 100 waves hit it per second, the frequence or pitch is 100. A vibrating tuning fork marked with the letter "G" gives off a frequency of 384 waves per second.

Here is how scientists represent high and low frequency sound waves.

LOW FREQUENCY WAVE   **HIGH FREQUENCY WAVE**

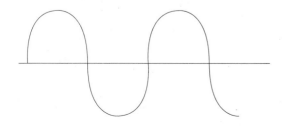

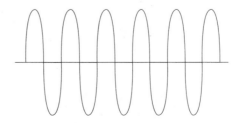

Name _____

Each sound has its own **loudness**. Loudness of a sound wave is roughly equal to the amount of energy it contains. Your ears can tell the difference between a loud and soft sound.

Here is how scientists represent loud and soft sound waves.

SOFT SOUND WAVE

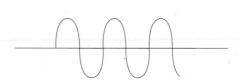

LOUD SOUND WAVE

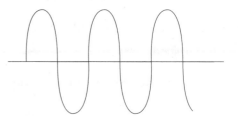

How Sounds Differ

Now let's repeat the ruler extending over the table experiment.

1. Hold the ruler so 8" (20 cm) extends over the table.

2. Hold the extended end of the ruler down a **small** amount and let it go. Listen carefully.

3. Bend the ruler again **to its limit** and let it go. Listen carefully.

In both cases you heard the same pitch or frequency. The second trial was louder because you gave the sound more energy.

4. Extend the ruler over 8", 6" and 4" (20, 15 and 10 cm) and repeat the experiment. This time pull down the extended part of the ruler the **same** amount each time.

In all three cases you had the same loudness. The frequency, however, changed with each trial.

At what inches (centimeters) was the sound the highest in frequency? _____

16

Tuning Fork Lab

What Is a Tuning Fork?

Tuning forks are U-shaped metal objects with a handle. They can be made of steel or aluminum. A tuning fork can give you a standard tone or pitch. Musicians use them to tune their instruments.

Most tuning forks have their pitch (also called frequency) stamped right on the handle. A tuning fork marked 256 gives off 256 vibrations per second. That is equal to "C" on the musical scale. A tuning fork marked 384 is equal to "G" on the musical scale.

You are going to experiment with tuning forks. Here are some rules to protect both you and the tuning forks.

**One-hole
rubber stopper**

1. Keep the vibrating U-shaped end away from glass, your eyes or your teeth.

2. Use the stem to hold the tuning fork at all times.

3. Strike the tuning fork **only** with the rubber hammer provided or on the side of your shoe. Never strike the tuning fork against metal or wood.

Tuning Fork Experiments

1. Obtain a tuning fork and a rubber hammer. You can make a rubber hammer with a pencil and a one-hole rubber stopper.

2. Strike your tuning fork. Twist it around near your ear.

Describe the sound. _____

3. Strike the tuning fork and place the vibrating end on your nose. **Not near your eyes**.

Describe how it felt. _____

Name _____

More Tuning Fork Experiments

1. Tape a light piece of thread to a Ping-Pong™ ball.

2. Strike and hold the tuning fork so it barely touches the suspended Ping-Pong™ ball.

Describe the amazing result. _____

_____

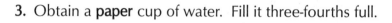

3. Obtain a **paper** cup of water. Fill it three-fourths full.

4. Strike and lower your tuning fork into the cup of water.

Describe your results. _____

5. Strike and place your fork **stem down** on a table.

Describe your results. _____

Try placing the stem down on other objects in the room. Try the wall, a trash can, the floor and even your head. Stay away from glass.

Newton Note: You noticed how the tuning fork vibration sounded louder and richer when touching a wooden table. The strings of a violin also vibrate like our tuning fork. It is the type of wood and the craftsmanship that separates a good violin from a cheap violin.

Over 300 years ago an Italian craftsman called Antonio Stradivarius made the world's finest violins. Some Stradivarius violins are worth millions today.

Tuning Fork Freedom

Here is a chance for you to develop two tuning fork experiments on your own. Remember to follow the safety rules listed in "What Is a Tuning Fork?"

18

Name _____

Freedom Experiment 1

1. What do you plan to do? _____

   _____

2. Sketch your experiment in the box below.

   +---+
   | |
   | |
   | |
   | |
   | |
   | |
   +---+

3. What were your results? _____

   _____

Freedom Experiment 2

1. What do you plan to do? _____

   _____

2. Sketch your experiment in the box below.

   +---+
   | |
   | |
   | |
   | |
   | |
   | |
   +---+

3. What were your results? _____

   _____

Name _____

A World Without Sound

Sound Is Your Friend

Imagine a world without sound. Imagine never hearing a bird singing.

1. What would you miss the most in a world without sound? _____

_____

2. How would a world without sound be dangerous to humans?

_____

Sign Language Alphabet

Sign language is a system of hand signs and gestures. It is a means of communication for hearing-impaired people. Primitive man used sign language. Some South American Indians still use a form of sign language. Below is the sign language for the letters of the alphabet.

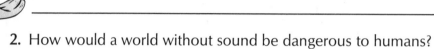

| A | B | C | D | E | F | G | H | I |

| J | K | L | M | N | O | P | Q | R |

| S | T | U | V | W | X | Y | Z |

Study the alphabet. Try signing the following words: dog, cat, home, radio, movie, your first name.

Name _____

Pair off.

Try to send and receive some simple words.

Feel free to use the chart as much as you need.

Signing Words or Ideas

There are also sign language symbols for words and ideas. Below are a few of these symbols.

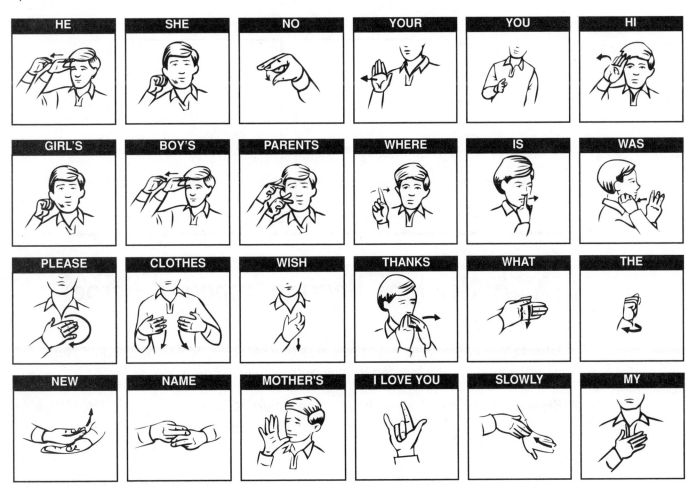

1. Study the chart above.

2. Practice some of the hand signals by working with a partner. See if he or she can recognize your hand symbols. Feel free to use the chart.

3. Use the chart to try to sign a simple sentence to your partner. Reverse roles and try to receive a simple sentence.

Name _____

4. There are obviously more symbols than in our simple chart. Try to create a hand symbol for the following ideas. Draw your hand signals in the boxes below the words.

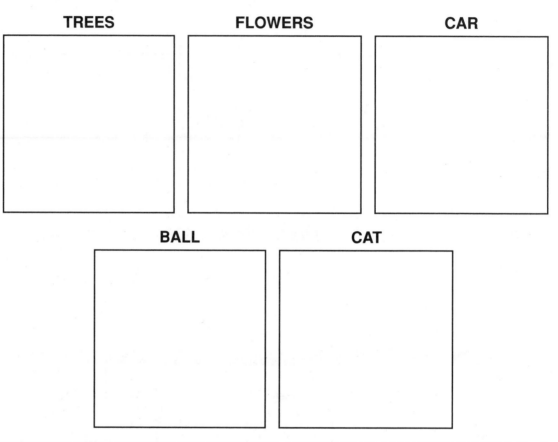

TREES

FLOWERS

CAR

BALL

CAT

Newton's Hearing-Impaired Heroes

Ludwig Van Beethoven: Beethoven was one of the world's greatest composers. At a very young age, he learned to play both the violin and piano.

Beethoven began to lose his hearing in his thirties. He became totally deaf during the last years of his life. Even with that handicap, Beethoven continued to write inspiring music.

Helen Keller: A serious illness at age two caused Helen Keller to lose both her sight and hearing. She couldn't even speak.

At age seven she was taught how to spell out words using her sense of touch. She finally learned to speak well enough to go to college. Helen Keller even learned to "listen" to other people by placing her hands on their lips.

Helen Keller worked all her life to help people with handicaps.

Name _____

Thomas Alva Edison: Edison was probably the world's greatest inventor. He patented over 1100 inventions during his lifetime. We owe the phonograph, movies and the incandescent light bulb to this genius.

Edison became partially deaf early in life. His deafness became worse as he grew older. He told friends that he liked not hearing too well. The deafness allowed him to concentrate on his inventions.

Try to research a famous person who was hearing-impaired. He or she doesn't have to be a scientist. Share your research with others.

Name _____

The Math of Sound

How Fast Does Sound Travel?

Sound travels through solids, liquids and gases. Sound cannot travel in a vacuum. The moon has no air, therefore, there cannot be sound on the moon.

Sound travels at these speeds in **air**.
   1100 feet per second
    330 meters per second

Sound travels at these speeds in **water**.
   5000 feet per second
   1500 meters per second

Sound travels at these speeds in **steel**.
  16,500 feet per second
   5150 meters per second

The speed of sound in air was given for a temperature of 32°F or 0°C. The hotter the air, the faster sound travels. You can calculate the speed of sound in air using the formula below. Here is an example for Fahrenheit temperatures only.

The temperature is 50°F.

Speed of sound in air = 1100 + 1.2 T (temperature)
                      = 1100 + (1.2 x 50) =
                      = 1100 + 60 = 1160 feet per second

1. Compute the speed of sound in air for a 100°F temperature.

   _____ feet per second

2. Compute the speed of sound in air for your classroom temperature.

   _____ feet per second

3. Why is there no sound on the moon? _____

4. Does sound travel faster in air, water or steel? _____

Name _____

Sound Versus Light

Sound travels 1100 feet per second. Light travels 186,000 miles per second.

Sound travels roughly one-fifth mile in one second. Light is so fast that it could travel almost eight times around the Earth in the same one second.

This information about the speed of sound and light can be very useful during a thunderstorm. Let's say that you see the flash of lightning that is one mile away. The light travels so fast that you see it almost instantly. Sound travels only one-fifth of a mile a second. You wouldn't hear the thunder until five seconds **after** you saw the lightning.

1. Lightning struck four miles away. How many seconds will it be before you hear

   the thunder? _____

2. You see lightning and start counting the seconds until you hear thunder. Ten sec-

   onds go by and you hear the thunder. How far away was the lightning? _____

Next time you are in a lightning storm, try estimating its distance. As soon as you see lightning, start counting the seconds till you hear thunder. Every five seconds is equal to a mile.

> **Newton Hint:** Lightning is caused by an electric spark. The spark can be from cloud to cloud or cloud to ground.
>
> Thunder is due to the lightning heating the air in its path. The air expands violently causing the sound of thunder.

You Are Hurting Newton's Ears

Have your mom and dad complained about your loud music? They are right to complain. Excessively loud sounds, such as your loud music, can cause headaches, irritability and anxiety. So turn the music down and improve your parents' mood.

Sound loudness is rated in units called **decibels**. The higher the decibels, the louder the sound. Distance from the origin of sounds reduces the decibels that your ears receive. Notice the chart on the next page gives a distance along with each decibel rating.

Name _____

Whisper	Light traffic	Jackhammer	Loud shout	Rock concert	Jet airplane taking off

| 30 decibels at 15 feet | 55 decibels at 50 feet | 85 decibels at 50 feet | 100 decibels at 50 feet | 117 decibels at 100 feet | 140 decibels at 200 feet |

1. How many decibels would you guess for a jet engine taking off 50 feet from you? _____ decibels.

Use the chart to **estimate** the possible decibels for each situation below.

2. Home noises _____ decibels.

3. Normal conversation (1 foot) _____ decibels.

4. Heavy traffic (50 feet) _____ decibels.

5. Fire engine siren (30 feet) _____ decibels.

6. Survey your home. List the loudest and quietest areas. _____

_____

The Sound of Voices

Birds sing and lions roar. Cats meow and dogs bark. All these animals and you use vocal cords to make sounds. Bees and flies make buzzing sounds through their wing vibrations.

The human voice is remarkable because of its wide range of sounds. The pitch or frequency of your voice depends mainly on your two vocal cords. As children, both boys and girls have similar pitches. As boys go through puberty, their vocal cords get longer and their voices get deeper. Here's a simple experiment to show why boys have deeper voices.

1. Obtain two similar rubber bands.

2. Cut them both in **one** place. Now you have two single bands of the **same** length.

3. Cut **one** of the rubber bands in half. Let the long rubber band represent the mature boy's vocal cords. Let the short rubber band represent the mature girl's vocal cords.

Name _____

4. Hold each with the **same** tension. Pluck them both so that they vibrate and give off sound. Try this a few times.

Which rubber band gave off the lower pitch? _____

Why do mature women's voices sound higher pitched than men's? _____

_____

ANIMAL VOICES AND THEIR AVERAGE PITCH

Human . . . . .600 cycles per second

Frog . . . . .1000 cycles per second

Cat . . . . .2000 cycles per second

Bird . . . . .10,000 cycles per second

Bat . . . . .50,000 cycles per second

Name _____

Sound Resonance

Newton Explains Resonance

Resonance happens when a small repeated push causes a large vibration. The small push must happen at the **right time** if there is to be large vibration.

Here is an example of resonance. A child is on a swing. An adult gives the swing a push. The adult times the push to just when the swing is moving forward. If timed correctly, the many small pushes will cause a large vibration in the swing.

Newton Note: Soldiers marching across a small bridge are told **not** to march in lockstep. The small force of many feet striking the bridge at the **same** time could cause the bridge to resonate and, perhaps, break.

Sounds That Resonate

Resonance is important in making musical instruments sound louder and better. Let's look at a sound wave drawing to understand how resonance happens.

HOW A SOUND WAVE LOOKS

A

|— 1 wave —|

The space between each wave is shown by the dots. One sound wave is a movement upward, downward and back to the next dot. The wave is moving from left to right.

28

Name _____

1. How many sound waves are shown in the diagram on page 28? _____
   Assume the waves you counted are moving past point "A" in one second. The number of waves passing a point in one second is called its **frequency**.

2. What is the frequency for the sound wave shown? _____ cycles per second

Tuning Fork Resonance

Tuning forks can make a pleasant sound. Here's an experiment to use resonance to make your tuning fork louder and more pleasant.

1. Obtain a tuning fork and a rubber mallet to strike it with.

2. Obtain a plastic or glass tube about 12" (30 cm) long and at least 1" (2.5 cm) in diameter.

3. Place the tube in a large container of water.

4. Strike the tuning fork and place it just above, but not touching, the tube.

5. Move the tube up or down while the tuning fork is vibrating. At one point you will hear a very loud sound. This is resonance. Measure the distance

   from the top of the tube to the water surface. _____ inches (centimeters)

Newton Explanation: Your tuning fork gives off a certain amount of sound waves per second. At the point of resonance, the sound wave has just enough time to bounce off the water's surface. Its echo reinforces the tuning fork as it begins to move upward. This is the same as timed pushes on a swing. Both the swing and your tuning fork vibrated more strongly because of resonance.

Resonance Math

Your tuning fork probably has a number on it. The number could be 128, 384 or 480. This tells you the number of vibrations it makes in cycles per second.

You can roughly check out a tuning fork's frequency with this formula.

$$F = \frac{1100}{W}$$

1 wave

F = frequency in cycles per second

1100 is the speed of sound

W = wave length of the sound wave

Name _____

In the "Tuning Fork Resonance" section, you measured the length of the tube at the point of resonance. This turns out to be one-fourth of the sound's wave length.

1. Multiply the tube length in the "Tuning Fork Resonance" section by 4. This gives you your tuning fork's wave length. _____

2. Use the formula above to divide your wave length into 1100 (the speed of sound). The answer is your tuning fork's frequency. _____

3. Compare your math results with the actual tuning fork frequency. You will usually be a bit lower.

4. Repeat the "Tuning Fork Resonance" and "Resonance Math" sections with different tuning forks.

The Sound of Music

Musical Instruments

You like music. Your parents and grandparents also like music. Different generations may have distinctly different tastes in music. Yet they all agree on the pleasure music gives them.

Musical sounds are made by three different kinds of instruments.

Stringed Instruments: Violins, guitars and pianos. Their strings vibrate against a box or board. You can change the frequency of a string by making it tighter, shorter or thinner.

Wind Instruments: Flutes, trumpets and trombones. A stream of air blowing across an opening in a tube or pipe starts the sound. The frequency of wind instruments is changed by lengthening or shortening the air column. A musician's vibrating lips can also affect the sound of a wind instrument.

Percussion Instruments: Bells, drums and cymbals. The vibration of the particular drum skin or bell metal affects the sound.

Anyone can buy a musical instrument. In this lab you are going to make your own.

A Musical Competition

You are being asked to make simple homemade musical instruments. You will form teams to make **two** different kinds. Your team will be given 10 days to plan and build your instruments.

Some suggestions are given on the next page. They are only rough ideas. It is your job to build them bigger and better.

Here are some ways in which your instruments will be judged.

1. Can you demonstrate a high-pitched and low-pitched note?

2. Can you play a musical scale?

Name _____

3. Can you play a tune that the class can recognize?

4. Can you use simple, inexpensive materials?

5. Can you use two of your instruments together to create a band effect?

6. Can you construct your instruments so that they look, as well as sound, beautiful?

7. Can you come up with a novel instrument different in some ways from the instruments shown below?

8. Can you and others enjoy the music you make?

Instrument Suggestions

Feel free to find other instruments from other sources.

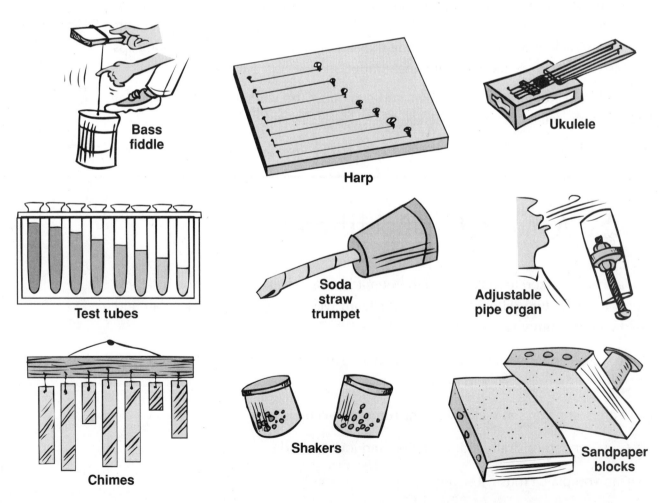

Bass fiddle

Harp

Ukulele

Test tubes

Soda straw trumpet

Adjustable pipe organ

Chimes

Shakers

Sandpaper blocks

Hearing Sound

Your Ears at Work

Your ears are your second most important sense. Only your eyes give you more information about the environment.

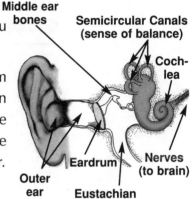

Your outer ear helps collect sound vibrations. These vibrations, in the form of air molecules, move your eardrum. The eardrum passes the vibrations on to three tiny bones that amplify the sound. The vibrations then affect the fluid in a snail-shaped chamber called the **cochlea**. From the cochlea, nerve cells carry impulses to the brain. The brain interprets the sounds you hear.

This lab will be a challenge to your ears.

Sound Direction

You have two ears. They are duplicates of each other. Each ear collects sounds and passes them on to the brain.

So why do you have two ears? Two ears help you determine the **direction** of sound. Let's experiment.

1. Obtain anything that clicks. It could be a retractable pen or anything similar.

2. Blindfold a volunteer. Have him or her cover **one** ear firmly with their hand.

3. Click 10 times from 10 different positions behind, in front, to the side, above or below your volunteer's head.

4. How many of the 10 clicks did the volunteer with one ear correctly locate?

   _____

5. Keep the blindfold on and repeat the experiment without any hand covering an ear.

6. How many of the 10 clicks did the volunteer correctly locate with **both** ears?

   _____

Name _____

7. Repeat the experiment using other volunteers. Scientists always repeat their experiments.

8. Look over your experiment results.

What do they prove about one- and two-ear sound location? _____

_____

Newton Knowledge: Other animals have ears. A rabbit's outer ear is very large. Large ears can collect more sound waves and warn of danger. Rabbits even adjust their outer ears to pinpoint the exact direction of sound.

Hearing Distance

Sounds get weaker the further away you are from them. Here is a simple way to check your ability to hear distant sounds.

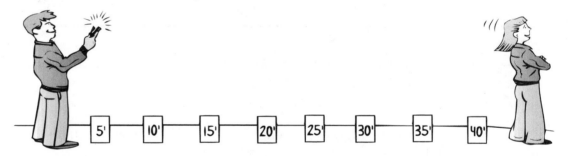

1. Use the same clicker as in the "Sound Direction" section.

2. Mark off 5' (1.5 m) distances on the floor as shown in the diagram.

3. Have one person stand at 0'. He or she will send out clicks in groups of twos, threes or fours.

4. Have the tested student stand at the 10' (3 m) distance and look **away** from the clicker.

5. The clicker will click two, three or four times at random.

6. Each time the tested student correctly identifies the number of clicks, he or she moves five more feet (1.5 or more m) away.

7. Continue moving the tested student until he or she misses the number of clicks.

8. Record the results for all the students being tested in the Sound Distance Data Table on the following page.

Name _____

SOUND DISTANCE DATA TABLE	
Student Name	**Last Distance Heard in Feet (Meters)**
Example: Bill	30

9. Repeat the preceding steps for all the students on your team.

Which student had the best distance hearing? _____

10. Check your data table. Did boys or girls have the better distance hearing?

_____

Hearing Freedom

Extremely loud noises can hurt your ears. What materials work best to muffle sound and save your ears?

Newton wants you to work out the details of this hearing test. You will obviously need a source of loud sound such as a radio, bell or a "?". You will also need materials such as metal, glass, cardboard, wood, plastic, cork, water or "?" to try to muffle the sound.

A good sound experiment will need a loud sound that stays the same loudness. Your materials should have the same thickness for each. You will also need "standard student ears." Perhaps you can use the best ear you found in the "Hearing Distance" section as your "standard student ears."

You may not be able to create the world's greatest hearing experiment. Just do the best you can. Try to organize your results in a data table.

Newton Note: In this last experiment you are working as **acoustic** engineers. Acoustics is the science of sound. Acoustical engineers use various materials to control sound in cars, refrigerators and auditoriums.

Name _____

Sound Science Frontiers

Sound Is for More Than Hearing

Scientists have found many uses for sound waves. Normal sound waves that you hear vibrate between 20 and 20,000 times per second. Sound waves higher than 20,000 cycles per second are called **ultrasonic**. They can have frequencies that run into millions of vibrations per second.

Ultrasonics can be used for testing airplane parts and welding plastics. It can be used to observe unborn babies. Ultrasonics underwater can detect schools of fish or an enemy submarine. It can be used to pasteurize milk or sterilize instruments.

You probably never realized how useful sound waves can be. Newton wants you to educate yourself and your classmates on sound science frontiers. Here is your sound science frontier research assignment.

1. Pick a sound science topic from the list below or find any sound science topic on your own.

2. Prepare a two- to four-page report on your sound topic. Drawings and photos are encouraged.

3. Make a simple non-working model of the device or idea you have chosen.

4. Be prepared to have both a written and oral presentation ready on _____.

SUGGESTED SOUND SCIENCE TOPICS

Acoustics	Fathometer	Microphones	Stereophonics
Animal Sounds	Hearing Aids	Mufflers (auto)	Stethoscopes
Bats and Echoes	Human Voice	Noise Pollution	Telephones
Deafness	Human Ear	Sonar	Any sound topic of your choice
Doppler Effect	Mach Number	Sound Recording	

Mirror Science

Newton Explains Mirrors

Mirrors are very useful. You can use them to admire yourself. You can mount them on cars to improve visibility. Mirrors in telescopes enable you to observe distant stars.

Any **smooth** surface can act as a mirror. Here is how to use water's surface as a mirror.

1. Tape some black paper to the bottom of a shallow bowl.

2. Fill the bowl half full with water.

3. Look straight down at the water's surface to see your reflected self.

4. Stir the water surface with your finger while trying to view yourself as before.

5. What happened to your reflected image? _____

The water surface makes a poor mirror. The best mirrors are made with flat panes of glass. One side of the glass is coated with both silver and black paint so that light cannot pass through. Instead the light is reflected back to you.

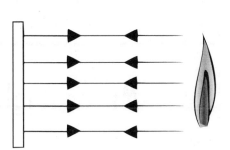

The diagram to the right shows how light rays hit and are reflected back by a flat mirror.

Strange Mirrors

1. Look at yourself in a mirror.

2. Wink your left eye.

3. Which eye seems to be blinking in the mirror? _____

Name _____

You just discovered that a mirror reverses an image from right to left. Mirrors do not reverse an image from top to bottom.

WIRROR
MIRROR

4. Touch your head with your hand.

5. Is your head up or down in the mirror? _____

6. Place a mirror as shown to the left with the word *mirror* printed in large print.

7. Is the M upside down? _____

8. Is the M image on the right or left? _____

Upright mirror

Ruler

What you see in a mirror is called a **virtual** image. It is called *virtual* because it is not really there.

9. Stand a mirror on its **edge** on a table.

10. Place a ruler on the table at a right angle to the mirror.

11. Place a penny at the 3" (8 cm) mark. Estimate how far behind the mirror that the penny's virtual image seems to appear. _____ inches (cm)

12. Repeat at 6" (15 cm) and give your estimate as to how far behind the mirror the penny seems to be. _____ inches (cm)

13. Repeat for 12" (30 cm). _____ inches (cm)

You have learned that images in a mirror appear to be the **same** distance **behind** the mirror as they really are in **front** of the mirror.

Mirror Angles

Light rays reflecting from a flat mirror obey the **law of reflection**. The angle at which the light hits the mirror is the same angle as the reflected light.

 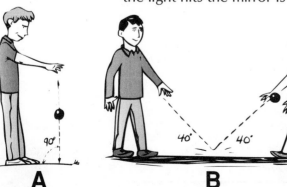

You can use a ball to demonstrate light reflection. Bounce a ball straight downward as in illustration A. The ball bounces right back to you. The ball went down at a 90° angle. It bounced back at a 90° angle.

Now bounce the ball to a friend as shown in illustration B. The ball returns to your friend at exactly the angle that you threw it.

A   **B**

38

Name _____

Light rays act like your bouncing ball. They bounce off a flat mirror at the exact angle that they appear.

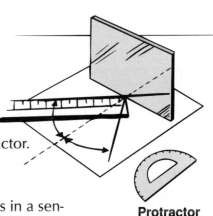

Mirror

Here is a more exact way to measure light ray angles.

1. Obtain a mirror, ruler and a protractor.

2. Draw a **dotted** line on a piece of paper.

3. Draw a solid line joining it and any angles **less** than 90°.

4. Place an upright mirror as shown.

5. Line up a ruler with the **reflection** of the solid line as shown.

6. Use the ruler to draw the solid line that you see.

7. Measure the original angle and the reflected angle with the protractor.

   _____° original angle   _____° reflected angle

Protractor

8. Sum up what you have learned about light rays, mirrors and angles in a sentence. _____

   _____

 # Curved Mirrors

There are two kinds of curved mirrors. A mirror that curves outward is called **convex**. A mirror that curves inward is called **concave**. The diagram to the right shows how a beam of light is reflected from each kind of mirror.

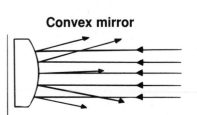

Convex mirror

You will find concave mirrors in automobile headlights. They focus light from the bulb into a straight beam of light.

You will find a convex mirror used as the passenger side mirror in automobiles. They give the driver a larger than normal field of view.

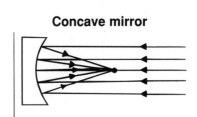

Concave mirror

1. Obtain a flat mirror, convex mirror and a concave mirror. You could also use the inside and outside of a shiny spoon for the concave and convex mirrors.

2. Observe yourself in both a flat and a **convex** mirror.

Describe the differences in the reflected image. _____

_____

Name _____

3. Move both the flat and convex mirrors to and from your face.

Describe the differences in the reflected image.

_____

_____

4. Now observe yourself in both flat and **concave** mirrors.

Describe the differences in the reflected image.

_____

_____

5. Move both the flat and concave mirrors to and from your face.

Describe the differences in the reflected image.

_____

_____

Fun with Mirrors

Mirror Multiples

Mirrors are interesting. They reverse images from left to right but do not reverse up and down. You always appear to be as far behind a mirror as you actually are in front.

Mirrors can multiply. They can't do multiplication tables, but they can give multiple images.

1. Place a pushpin into a piece of corrugated cardboard.

2. Place two mirrors as shown behind the pushpin.

3. Observe at eye level as you move the mirrors together.

Newton feels left out. Place your two mirrors behind Newton. Move the mirrors together and multiply Newton.

Mirror Words

Letters and words often appear strange when viewed in a mirror. Some letters and words look normal in a mirror.

1. Place a mirror at the dotted line above the word *cook*.

2. Does it appear normal in the mirror? _____

3. Place your mirror above the words *carbon dioxide* and *dry ice*.

4. Which **words** appear normal and which do not? _____

5. Which **letters** appear normal and which do not? _____

COOK

CARBON
DIOXIDE

DRY ICE

Name _____

6. There are 26 letters in our alphabet. Using uppercase letters, only nine look correct in a mirror. Can you list all nine? _____

7. *DIKE* and *BIKE* are mirror words. Can you list six more mirror words? _____

_____

Mirror Writing

1. Try writing (not printing) your name while viewing the process in a mirror. Some people find this difficult.

Isaac Newton

2. Isaac Newton's signature is on the left. Place a mirror above it. Now try to copy exactly what you see in the mirror. If you have done a good job, you should see Mr. Newton's correct signature viewed in a mirror.

3. Have some fun with mirror writing. Write a **short** message to your friend. See if he or she can identify the messages before viewing with a mirror.

Periscope Fun

A periscope is made of two flat mirrors. Try to position your mirrors to do the following:

1. Observe the back of your head.

2. Look at someone behind you.

3. Spy on a friend from around a corner.

A good periscope has the two mirrors in a box that keep out unwanted light rays. Tape two milk cartons together. Insert two mirrors at the same angle. Cut an opening for your eye and for the top mirror. **Go spy**!

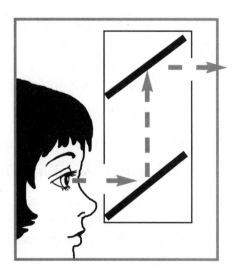

42

Name _____

Lens Science

Newton Explains Refraction

Cameras, telescopes and eyeglasses all use glass or plastic lenses. Lenses are used to bend light rays so that they either come together or separate.

The bending of light as it passes from air to another substance is called **refraction**. Refraction occurs because light slows down as it moves from air through glass, water and other substances.

The speed of light is 186,000 miles (300,000 km) per second. Light is so fast that it could travel almost eight times around the Earth in the same one second.

SPEED OF LIGHT IN DIFFERENT SUBSTANCES
(IN KILOMETERS PER SECOND)

Air 300,000
Water 225,000
Glass 200,000
Diamond 125,000

When light moves from air to water or glass, it slows down. The slowing down causes the light rays to bend. Here are some ways to see the refraction (bending) of light for yourself.

1. Obtain a glass of water, a pencil and a ruler.

2. Place your thumb under the water.

3. Place a pencil halfway into the water.

4. Place a ruler halfway into the water.

5. Observe each from different angles.

Describe how each appeared. _____

_____

Name _____

Lens Shapes

Concave lens

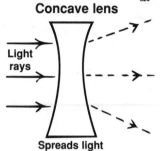

Light rays

Spreads light

Lenses are usually made of glass that has been ground and polished. There are two main types of lenses. **Concave** lenses spread light rays apart. **Convex** lenses bend light rays together.

1. Obtain a concave and a convex lens. Often an eye doctor can provide lenses. Magnifying glasses are usually convex. Eyeglasses can be convex or concave.

2. Use **both** kinds of lenses to view the print on this page. You will have to move the lenses up or down.

Describe the differences between the images with the concave and

convex lenses. _____

_____

Convex lens

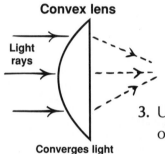

Light rays

Converges light

3. Use **both** lenses to view a distant object such as a tree out the window or an object on a far wall.

Describe the differences between the images of the two lenses. _____

_____

_____

4. Use **both** lenses to observe the face of a classmate.

Describe the differences between the images of the two lenses. _____

_____

_____

Measuring the Focal Length of a Lens

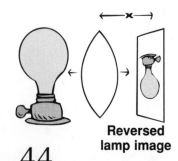

Reversed lamp image

A convex lens brings light together to form a focused image. Finding the distance between the lens and its sharp image helps you find its **focal length**.

Here's how to measure a lens' focal length.

1. Obtain a convex lens, a 40- to 60-watt light source and white cardboard.

2. Move the lens back and forth until you observe a **sharp** image of the bulb on the cardboard. The image will be upside down.

44

Name _____

3. Measure the distance "X" between the lens and the cardboard with the sharp image. _____ inches (centimeters)

The nature of a convex image is that it appears at **twice** the focal length.

4. Divide your "X" length by 2 to obtain the real focal length of your convex lens. _____ inches (centimeters)

5. Try to get the focal length of larger or smaller lenses.

Observing Refraction

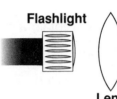

Flashlight

Lens

Aquarium or large jar

It is difficult to see light rays in transparent air. Let's try to observe light rays bending in milky water.

1. Obtain a concave and a convex lens, a strong flashlight, a small aquarium (or a large jar) and some nonfat milk.

2. Fill the aquarium with water.

3. Add enough nonfat milk to give the water a milky sheen. Don't use too much milk.

4. Set up the materials as shown. You may have to adjust the flashlight or lens distance.

5. Darken the room and observe the flashlight beam as it passes into the milky aquarium. Don't use the lenses yet.

6. Now place the convex lens between the flashlight and the aquarium. You may have to adjust the distances.

What happened to the light beam in the aquarium? _____

_____

7. Repeat using the concave lens.

What happened to the light beam in the aquarium? _____

_____

Newton worked with telescopes. Some telescopes are made with mirrors. The largest telescope made with a lens is at the Yerker's Observatory is Wisconsin. Its glass lens is 40" (102 cm) in diameter.

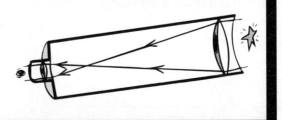

Fun with Lenses

Taking a Close-Up Look

Many things look different when looked at through a magnifying glass. Magnifiers use a convex lens.

1. Obtain a magnifying glass.

2. Obtain a collection of stamps, coins, dollar bills, flowers, leaves, leather (your wallet), colored magazine pictures, tissue, clothing, sponge, insects, chalk or "?". There should be many items available around you. Even try your fingerprints.

3. Look closely at the back side of a shiny penny. You are looking at the Lincoln Memorial.

   Whose image do you see **inside** the Memorial? _____

4. Look at a colored picture from a magazine or book.

   What are the colors really made of? _____

5. Now it's magnifier freedom time.

   List all the objects you observed. _____

   _____

   Which two were the most interesting? Explain. _____

   _____

On the right, draw and label the magnified item that looked the most interesting.

Name _____

Casting Images

Lenses can be used to focus light. Here are some images you can focus.

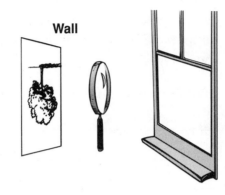

Wall

1. Darken the room except for a bright window.

2. Place you convex lens **near** a wall facing the window.

3. Move your lens back and forth until you get a sharp image of the window and the outdoors.

Describe what you saw on the wall. _____

_____

_____

4. Obtain a lamp. It should be 50 watts or better.

5. Darken the room except for the lamp.

6. Place a white card above the lamp with a lens just below it as shown.

7. Adjust the card or lens to focus an image of the lamp.

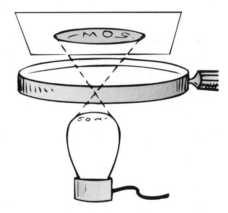

Were you able to read the wattage printed on

the top of the light bulb? _____

_____

_____

Name _____

Light Freedom

Sir Isaac Newton experimented with both lenses and mirrors. He discovered many things about the nature of light. Today your team can pick up where Newton stopped. Combine both **lenses** and **mirrors** in different ways to advance the science of light.

Describe what you did and the discoveries you made. _____

_____

_____

_____

_____

_____

Flashlight Science

How Does Light Travel?

Light travels in all directions from a light source. The light from our sun goes in many directions besides our planet Earth. A lamp at home sends light throughout a room.

Here is an experiment to discover a rule about light travel.

1. Obtain a flashlight and three cards that are about 5" x 8" (13 x 20 cm).

2. Cut a quarter-sized hole in the **center** of each card.

3. Make a right angle bend at the 2" (5 cm) mark.

4. Add two paper clips to the bent end of each card to add stability. See diagram.

5. Darken the room and **line up** the cards about 3" (8 cm) apart.

6. Shine a flashlight through the lined-up cards.

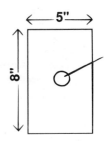

What do you see on the wall? _____

7. Now move the three cards so that the holes do **not** line up.

What do you see on the wall? _____

**Paper clips
on end of card**

8. You have discovered a light rule. Unscramble the sentence below that gives the rule.

    rays straight travel line in light a. _____

Flashlight Rainbow

White light from the sun or your flashlight contains all the colors of the rainbow. Rainbow colors always appear in the following sequence. They are red, orange, yellow, green, blue, indigo and violet.

White light splits into its rainbow colors when bending through glass or water. Water drops in the air actually cause the rainbow you sometimes see after a rainstorm. Following is an exercise to see how to use a glass of water to create a primitive rainbow.

Name _____

1/4" slit

1. Obtain a flashlight, black tape, a glass of water and a large white card.

2. Use the black tape to cover the front of the flashlight except for a 1/4" slit. It doesn't have to be exactly 1/4".

3. Place the glass of water on the edge of a chair. Place the white cardboard right below the chair.

4. Darken the room.

5. Adjust the flashlight beam so that it angles through the glass of water as shown.

6. Adjust the position of the cardboard on the floor so that it is centered on the floor so that it is centered on the rainbow that forms.

List the colors in the order that you see them. _____

_____

What happens when a white light beam is bent going through water or

glass? _____

_____

Light Through a Water Pipe

Fiber optics is relatively new. The optic fibers carry light in much the same way that copper wires carry electricity. Both the light fibers and the copper wire can carry messages over long distances. Both fibers and wires can be bent and twisted and still carry the light or electricity.

A stream of water can also carry a light wave. It does a poor job of moving light, but it can be done. Here is how to demonstrate a water light pipe.

Cover jar with black paper

Nail holes in lid

1. Obtain a tall thin pint jar. Olive jars work fine.

2. Cover the sides with black paper. Do **not** cover the top or bottom of the jar.

3. **Carefully** place two small nail holes in the lid. One hole should be at the center, the other at the edge as shown.

4. Fill the jar with water and place the lid on **tight**.

5. Darken the room.

Name _____

6. Pour the water out as shown over a sink while holding a
   flashlight to the bottom of the jar.

Describe what you see. _____

_____

_____

How is your water stream similar to fiber optics? _____

_____

_____

Flashlight Shadows

Wall

1. Darken the room.

2. Use a strong flashlight to cast a shadow of a ball (such as a tennis ball) on a wall.
   You may have to adjust the ball and flashlight distances to get a good shadow.

3. Study the ball's shadow.

Can you observe a dark shadow in the center and a lighter shadow surrounding

it? _____

Newton Help: Scientists call the dark central shadow the **umbra**. The lighter
surrounding shadow is called the **penumbra**.

Look up *solar eclipse* in a reference source. Find out if the moon throws two shadows on the Earth during a solar eclipse.

Name _____

Light and Your Eyes

Understanding Your Eyes

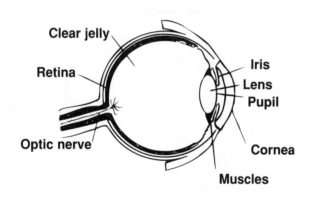

To the left is a diagram of your eye. Here are the major parts.

Iris: Colored part of your eye that adjusts the amount of light that can enter.

Lens: Adjusts its shape to focus light.

Pupil: Hole through which light enters the eye.

Cornea: Plastic window that protects the eye.

Muscles: Change shape of lens to focus light.

Optic Nerve: Carries nerve signals from the eye to the brain.

Retina: Converts light into nerve signals.

Clear Jelly: Fills the eyeball to keep its shape.

1. Which **two** parts of your eye can adjust themselves? _____

2. Your eyelids protect your eyes. What protects your eyes when the eyelids are up? _____

3. What turns light into nerve messages? _____

Eyes That Need Help

The lens of a normal eye adjusts to focus a sharp image on the retina. Some eyes cannot adjust the lens to give a sharp image. People with this problem see objects blurred rather than clear. Their eyeballs may be too long or too short.

52

Name _____

People who are **nearsighted** cannot see distant objects clearly. Their eyes focus light in front of their retina. They need glasses with a **concave** lens to spread light out.

People who are **farsighted** cannot see close objects clearly. The eye lens tends to focus light behind the retina. They need a **convex** lens to bend light inward.

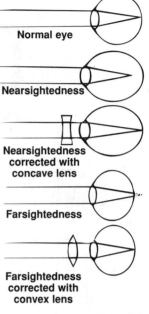

Normal eye

Nearsightedness

Nearsightedness corrected with concave lens

Farsightedness

Farsightedness corrected with convex lens

Many of you wear glasses. Glasses are lenses designed to bend the light entering your eye. Some glasses are convex and bend light inward. Some glasses are concave and bend light outward. Here is a simple way to find out if you are wearing convex or concave lenses in your eyeglasses.

1. Obtain a comb and a flashlight.

2. Place the comb **teeth down** a few inches (centimeters) in front of the flashlight.

3. Turn the flashlight on and darken the room.

Describe what you see beyond the comb. _____

4. Place a pair of eyeglasses in front of the comb as shown. Use only **one** lens.

What happened to the light rays as they went through the lens? _____

_____

Newton Help: If the glass was concave, the lines would be bent outward. If the glass was convex, the lines would be bent inward.

Was the lens you tested convex or concave? _____

5. Try some other eyeglasses to determine how they bend light rays.

Persistence of Vision

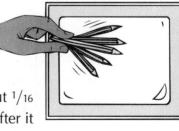

Images focus on your retina. Each image remains on your retina for about $1/16$ of a second. This means that you still see an image for $1/16$ of a second after it is no longer there. Here's a simple way to check your eye's persistence of vision.

1. Darken a room and turn on a television set or computer monitor.

2. Wiggle a pen or pencil rapidly in front of the screen.

How many pencils did you see on the screen? _____

Name _____

> **Newton Hint:** Your television draws 25 complete pictures each second. Each new picture leaves an impression of the pen or pencil.

Here's a more complicated method of demonstrating persistence of vision. Both television and motion pictures fool your eyes because of this persistence.

1. Obtain a 3" x 5" file card.

2. Newton wants you to make some choices on this experiment. You are basically going to put an animal in a cage using rotation. You will draw the animal of your choice on one side of the card. A suitable cage will be on the other side.

3. Draw an animal of your choice in one of the frames below. Use thick dark lines. Cut out the frame and place your animal on the center of the 3" x 5" card.

4. Cut out a cage for your animal from below. Place the cage on the center of the back of your 3" x 5" card.

5. Tape the card firmly on a pen or pencil.

6. Spin the pen or pencil **rapidly** between your hands and note the illusion due to persistence of vision.

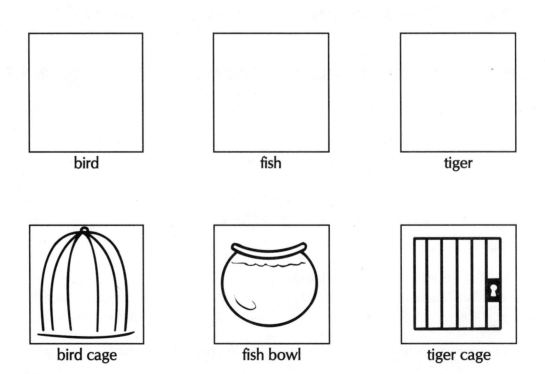

bird	fish	tiger

bird cage	fish bowl	tiger cage

Name _____

Light Waves Can Fool Your Eyes

Your eyes are marvelous instruments. They can detect objects miles away. Yet your eyes can easily be fooled. Here are some interesting optical illusions.

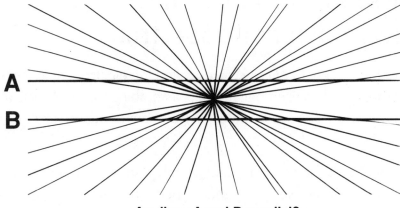

Are lines A and B parallel?

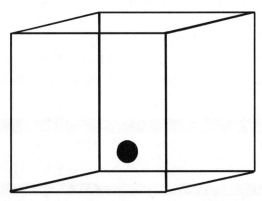

Stare at this staircase; then rotate the page until you spot the illusion.

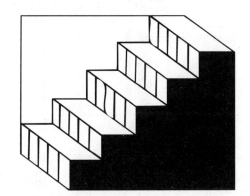

Is the dot at the back or front of the cube?

Name _____

The Spectrum of Light

Newton Introduces the Electromagnetic Spectrum

Light is only part of a variety of energy waves called the **electromagnetic spectrum**. The name *electromagnetic* is used because it is a combination of electric and magnetic energy. The word *spectrum* refers to the many kinds of electromagnetic energy.

Study the chart below. It shows visible light as only a small part of the electromagnetic spectrum. Most of the electromagnetic spectrum is invisible to us. We can only **see** light waves and **feel** heat waves.

WHAT IS IN THE ELECTROMAGNETIC SPECTRUM?

Gamma rays	X-rays	Ultraviolet radiation		Infrared radiation	Microwaves	Radio waves

Visible
Light

The longest energy waves are radio waves which activate our radios and televisions. Then there are microwaves that we use for cooking. Infrared rays are given off by hot objects such as our sun. The visible light spectrum contains all the colors of the rainbow that our eyes can see. Ultraviolet rays can tan us when given off by the sun or a sun lamp. The X-rays can penetrate our bodies and help doctors look inside. Gamma rays are the shortest in the electromagnetic spectrum. They are radioactive and dangerous.

All the waves in the electromagnetic spectrum travel at 186,000 miles (300,000 km) per second. Here are some examples of the time it takes these waves to travel various distances.

New York to London	0.1 seconds	Sun to Earth	8 minutes
Moon to Earth	1.3 seconds	Nearest Star to Earth	4.5 years

56

Name _____

Newton Discovers the Rainbow

Sir Isaac Newton discovered many important laws about motion and gravity. One of his most important discoveries was about the nature of light and color.

Red
Orange
Yellow
Green
Blue
Indigo
Violet

Newton invented a triangular piece of glass called a **prism**. He darkened a room and allowed sunlight to enter through a tiny slit. The prism bent this white sunlight into the colors of the rainbow. The colors of all rainbows come out in the order shown in the diagram. Use the name **Roy G Biv** to help you remember them.

1. Obtain a glass or plastic prism.

2. Darken the room and allow only a slit of light to enter.

3. You can also use a beam of light from a projector or a powerful flashlight.

Name the colors in their order in your rainbow. The color indigo is between

blue and violet. _____

Did you get more than one rainbow? Look carefully around the room. _____

> **Newton Note:** You may be wondering how nature makes a rainbow without prisms. Nature's prisms are really raindrops.

Building and Using a Spectroscope

There is another way of viewing rainbows. It uses a **diffraction grating** instead of prisms. Diffraction gratings are made of film with thousands of lines etched per square inch. As light passes through the spaces between lines, it is bent in much the same way as a prism.

1. Obtain a small shoe box (or anything similar), some black tape and a 1" (2.5 cm) square of diffraction grating.

2. Build the device shown at the top of page 58. It is called a **spectroscope**. Make the slit about 1/8" (0.3 cm) wide and 1" (2.5 cm) long. It is important that your slit has sharp edges.

The Spectrum of Light

Name _____

SHOE BOX SPECTROSCOPE

Top of box – close when complete

Light source

Eye

Diffraction grating– 1" square

¹/₈" slit made from black tape

Diffraction grating

20,000 lines per square inch

Your spectroscope can be used to see the spectrum (rainbow) from different light sources.   Observe carefully and you will see the differences.   Fill out the Spectroscope Data Table as you view various light sources. **Do not use your spectroscope to view the sun directly.**

SPECTROSCOPE DATA TABLE

Light Viewed	Colors Seen						Detailed Description
	red	orange	yellow	green	blue	violet	
sunlight–look out the window danger–do not look at sun							
fluorescent light (note the dark blue lines)							
incandescent light bulb (clear glass is best, but frosted is okay)							
neon, mercury or any gas light your teacher can obtain							

Name _____

Chemical Analysis with a Spectroscope

When any substance is heated to a high temperature, its atoms give off a special color. If viewed through a spectroscope, these colors are like fingerprints that identify the hot atoms. For example, if you know the "color fingerprints" of boron you can recognize it in the light from the sun or stars.

In this part of the lab **your teacher** will burn various chemicals. Students are to observe this experiment only. A teacher or a qualified adult must be the only one to handle or burn chemicals. First notice the color without a spectroscope, and then try to spread that color out into its fingerprint by viewing it through the spectroscope. Work fast, for it is difficult for your teacher to burn the chemical for a long time.

Chemical Burned	Color Viewed with Eye	Color Viewed with Spectroscope					
		red	orange	yellow	green	blue	violet
1. sodium (salt)							
2. zinc							
3. potassium (cream of tartar)							
4. boron (borax)							
5. iron filings (sprinkle on)							
6. strontium or lithium salts							
7. copper or copper sulfate							
8. magnesium ribbon–burn carefully							

Name _____

Light, Action, Camera

How Your Eye Is Like a Camera

The **eye's lens** focuses the image on the retina.

Retina

The **iris** opens and closes to allow the right amount of light to enter the eye so that we can see sharply and clearly.

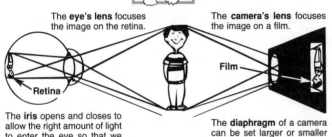

The **camera's lens** focuses the image on a film.

Film

The **diaphragm** of a camera can be set larger or smaller to allow enough light to fall on the film for a clear picture.

Study the diagram to the right. It shows how your eye is very similar to a camera. This will help you understand the simple pinhole camera that you will be constructing.

1. What part of your eye is like film in a camera?

    _____

2. What part of your eye is like the diaphragm of a camera?   _____

3. What focuses light in your eye and in your camera? _____

4. Notice the image on both the retina and film in the diagram. What is strange

    about both images? _____

Newton Note: Both your eye and the camera have an upside-down image due to light bending through the lenses. Your brain interprets the image correctly.

Building a Pinhole Camera

1. Obtain a small, clean coffee or dog food can. Try to find one with a clear plastic lid.

2. Hammer a **small** nail into the bottom of the can. Make the smallest hole that you can.

3. Cover the open end with the clear plastic lid. If you can't find a plastic lid, use wax paper. Stretch it neatly over the can and secure it with a rubber band. The wax paper must have a tight, smooth surface.

Name _____

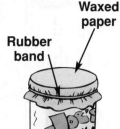

Rubber band **Waxed paper**

Coffee can

Using Your Pinhole Camera

Here is how to use your pinhole camera. Point the pinhole at the viewed object and observe the plastic lid or wax paper.

1. Darken the room and observe a candle flame. You may have just enough light to observe only the flame.

Is it right side up or upside down? _____

2. Darken the room and observe an open window.

Describe what you see. _____

_____

3. Darken the room. Observe a bright light bulb directly.

Describe what you see. _____

_____

Newton Note: Try to build a better pinhole camera. Try different size holes. Try other kinds of plastic or paper that are translucent. Try different size cans or cardboard boxes.

Candle

Window with tree outside

Bright light bulb

Name _____

Fun with Optical Illusions

 ## Optical Illusions

An *optical illusion* is defined as "something that fools our normal sense of perception." Below are three classical optical illusions for you to enjoy.

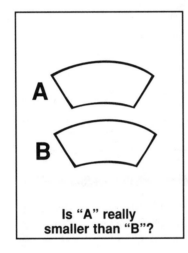

Is "A" really smaller than "B"?

Is the hat really taller than it is wide?

Are these men different in size?

 ## Making Money

Everybody likes to make money. Here is a way to convert two coins into three. Don't try to spend the third coin. It is an illusion.

1. Obtain two nickels or two quarters.

2. Hold them as shown between two fingers.

3. Place them about 6" (15 cm) in front of your eyes.

4. Slide the coins back and forth **rapidly** between your fingers until you "see" the third coin.

Coins at eye level

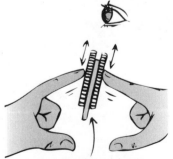

Third coin appears here

Name _____

The Square Circle Illusion

A square is a square. A circle is a circle. Here is a square that you can spin into a circle.

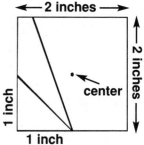

1. Cut a piece of cardboard into a 2" (5 cm) square.

2. Mark a dot exactly in the center.

3. Draw two dark lines approximately like the ones shown.

4. Push a pushpin through the center point as shown.

5. Use a small piece of tape to secure the pin to the cardboard.

6. Spin it on a tabletop.

   Describe what you see. _____

   _____

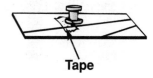

Tape

Answer Key

Newton's Sound Riddle, page 7

If you think of sound as a vibration, the falling tree gives off a sound.

Sounds That Resonate, page 29

1. Seven sound waves
2. Seven cycles per second

How Does Light Travel? page 49

Light rays travel in a straight line.

Understanding Your Eyes, page 52

1. Iris and lens
2. Cornea
3. Optic nerve